THE FORAGER'S GUIDE TO BOTANY

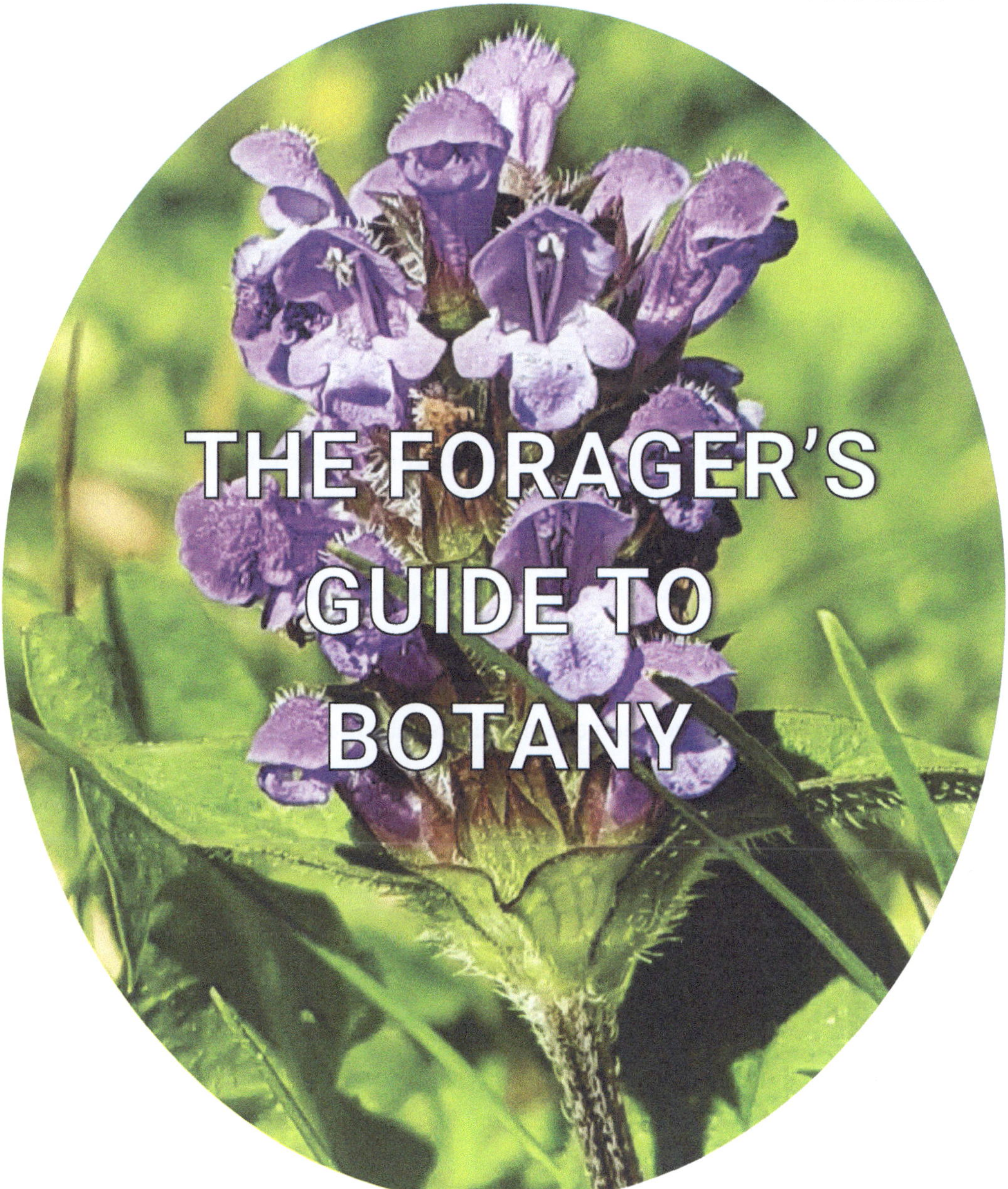

Other books from Gavin Ireland and Found Food

- *My Foraging Journal*

  This small notebook is perfect for foragers to record their finds, making sure that they can find them again in the future, and also to learn from experiences about environmental influences on places where specific plants/mushrooms are likely to grow.

Also available through foundfood.com

Since 2010, Found Food has been helping foraging enthusiasts to learn more about their surroundings and how they can best make use of the natural resources all around.

- *The Forager Helper*

  The Forager Helper is a repository of the knowledge that Gavin has built up over the years or foraging, wildcrafting, and studying. At the time of publishing there were over 100 plant, tree, and fungi monographs, videos, recipes, and plant and fungi family descriptions.

  Find out more at www.foragerhelper.foundfood.com

- *Simple Botany for Foragers*

  This is an online course which this book was designed to accompany. It includes video chapters, quizzes, and downloadable information sheets.

- *Face-to-face foraging walks*

  You can book face-to-face foraging walks covering Introductions to Foraging, Forage and Feast (which includes a foraging themed picnic), and Forage and Cook (which includes cooking a meal using the things we've foraged along the way), or you can request a custom walk/course.

- *Foraging Coaching*

  If you're looking to become a foraging teacher, or just want to develop your skills and deepen your understanding, a series of coaching calls can help you to achieve that.

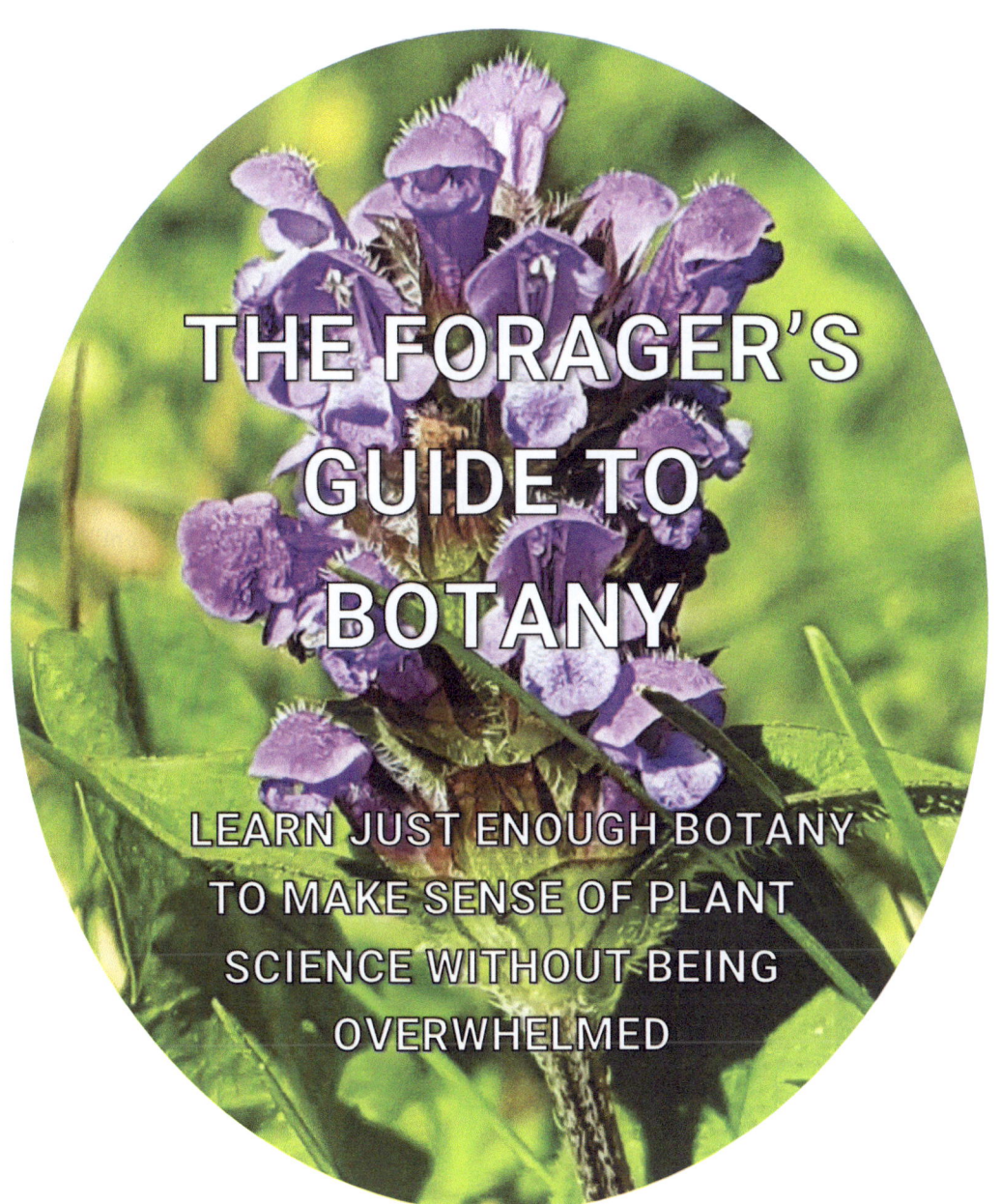

THE FORAGER'S GUIDE TO BOTANY

LEARN JUST ENOUGH BOTANY TO MAKE SENSE OF PLANT SCIENCE WITHOUT BEING OVERWHELMED

GAVIN IRELAND

TABLE OF CONTENTS

Found Food – The Forager's Guide to Botany

Why?

Why could a forager possibly need to know about botany?

Well, the easy answer is that most of the time you don't, but when you're trying to identify that frustrating plant and all the reference books are full of technical terms like 'stipe', 'axil', and 'bract', it helps to know what that all means. Or how about when you're looking for help online and five people are talking about 'Queen Anne's Lace' but they're each talking about a different plant? That's when understanding scientific names will help to cut through the confusion.

Botany is a degree course and a life-long learning curve though, isn't it?

Yes, it is; If you want to be a botanist, but it is possible to learn just enough botany to be a better forager and that's why I created this course and accompanying book. I've worked on understanding the minimum amount of botany needed to take your foraging journey to the next level, without getting bogged down into too much detail.

WHY HAVE I WRITTEN THIS BOOK?

I remember back when I was trying to advance my knowledge beyond the trees, plants and fungus that I had randomly learned about, because they were abundant near me, or interesting in some way, I started to invest in reference books, and keys, and to read the advice online from experienced foragers; and the thing that became clear very quickly was that there was whole world of terminology and structure that was alien to me. Where to start digging into this subject was not clear and there was always the danger that I end up looking into one subject area that ends up being of little use to a forager. Needless to say, I found a few of those subjects and fascinating though they were, they weren't a lot of use to my development as a forager.

I heard lots of hints and tips along the way, and suggested ways to progress my knowledge, but one theme kept coming up over and over again. So, here I make my confession: I did not invent the knowledge or the system that I'm proposing in this book. Many people already use it, some even train using it, but the difference here is that you'll find that it is specifically aimed at foragers, and I've aimed to avoid adding information you won't need. That's not to say that everything is about foraging, it isn't. Some of the concepts I want to introduce are easier to understand with a little background information. I've tried not to go too deep into botanical conventions and history though, so hopefully you'll find it useful.

The final, and possibly most important reason for this book's existence, is simply that I needed a system for myself and writing it down has helped to cement it into what I do; Because we're all still learning – No-one could ever know all there is to know about plants.

PART 1 – USING A SYSTEM TO IDENTIFY PLANTS

Using a system makes success repeatable, in just about all aspects of life. Random success will only get you so far. I spent many years accumulating knowledge of specific plants and trees which were either abundant near me or interesting in some way. Eventually I came to realise that whilst this way of learning plants wasn't terrible, sooner or later I would run out of memory capacity. There are almost 400,000 plant species currently identified, imagine trying to learn them all!

Memory has limited capacity for most people and the natural world seems almost limitless, so we need to use patterns and classifications. For example, some people's only concern about cars is that they can get you from A to B, whilst other people care whether they use unleaded petrol or diesel, and others may care about the exhaust emissions figures. The first group of people only care whether the car will get them to where they're going, so they only need to recognise a car. In the second group, those who prefer petrol need to learn to recognise cars which use petrol. In the final group, there will be people who need to recognise cars which use petrol and whose exhaust emissions are below a certain threshold.

Plants have classifications too, and at different levels of those classifications they can have different properties in common. For example, at the time of writing the Rose family, *Rosaceae*, has approximately 5000 species, many of them with edible parts. I honestly doubt that I could ever memorise the identification and characteristics of 5000 things! However, all members of the rose family have common features. If I can remember what those features are, then I've instantly whittled the possibilities down from 400000 to 5000, moreover if I can remember the one feature specific to the *Maleae* tribe within *Rosaceae*, I've now got around 1100 species worldwide whose fruit are all edible! (Some of the fruit isn't necessarily pleasant to eat, but none is poisonous). Some of the fruits of this family include apple, pear, quince, and rowan for example.

This system comes in a few parts, obviously it's your book so read it in any order you like, but I recommend:

1. Understand how plants names and classifications work, at a minimum Species, Genus, and Family.
2. Understand plant lifecycles – this will be essential in understanding what you might see from one year to the next.
3. Understand/learn the names for the various parts of a plant – so you know what the reference books are talking about.
4. Understand how to use plant keys, and practise, practise, practise.
5. Think about which plant families you see most where you live and try to learn the key characteristics of those families.

Remember, a system isn't a short-cut to knowing everything, it's just a way to make the journey a little easier.

CHAPTER 1 – PLANT NAMES AND STRUCTURE

Every plant indeed, every living thing that has been discovered, has a two-part name (binomial) based on a system invented by Carl Linnaeus in 1753. It is still sometimes referred to as the Linnaean classification. The names are based on Ancient Greek, Latin, and "Latinised" versions of modern language words.

Here's the first thing to not get hung up on – how to pronounce the scientific names. I have seen very intense arguments as to the pronunciation of a plant name, and some people have produced guides to pronunciation. However, I take a more pragmatic approach: Whilst some pronunciation can be assumed from modern languages derived from Ancient Greek and Latin, there is not one person on this planet who has ever heard Ancient Greek or Latin spoken aloud. Therefore, you may be confident in your pronunciation, but no-one can be 100% certain, therefore no-one can be proved to be pronouncing the names incorrectly.

BINOMIAL SCIENTIFIC NAMES

The first part of the plant name is the **genus** name (plural genera) and the first letter is always capitalised. The second part is the **species** name, and this is always lower case, and the whole name usually appears in *italics*. For example, *Glechoma hederacea*. Commonly known as ground ivy, creeping charlie, alehoof, field balm, tunhoof, and many others depending on who you ask and where in the world you are; However, the scientific name is only ever *Glechoma hederacea*.

Ground ivy (Glechoma hereacea)

Another way to look at this is to consider our way of naming ourselves. We could think of our surname as being our genus and our first name as being our species. I am species Gavin, of the genus Ireland. There are many members of the genus Ireland, but I am the only species Gavin in that genus.

Also, you can look at the derivation of the words; genus and species: genus = general, species = specific. Continuing the previous scenario, I am generally one of the Irelands, but specifically Gavin Ireland.

Genus Abbreviations

Back to plants now, it is not uncommon to see abbreviations in written scientific names. For example, if it has already been established that the author is writing about the *Glechoma* genus, they may choose to abbreviate to *G. hederacea, G. grandis, G. hirsuta* and so on.

The whole genus

If you see *Glechoma spp.* this means that the author is referring to all of the species contained within the *Glechoma* genus.

Relationships

Now here is where learning can begin to accelerate. All of the plants within the genus Glechoma will share certain characteristics, some of which will be things that we can see, feel or smell, which means that without going into details, instead of one specific plant, we can now confidently identify a group of related plants. However, let's just wait a second there. At the time of writing, there were 236 different genera within the *Lamiaceae* family. It's not going to be practical to learn the characteristics of 236 genera of just one family, not when there are over 600 families! Fortunately for us, plant families usually have enough common characteristics for us to focus at that level, but still over 600 right? Well, no. You won't find all 600 plus families growing in your local area, and some families are much more common and widespread than others, so that's where we start. For this book, I've included six common families to get started, but the system is to just start with one. Before you know it, you'll be saying "I'm not sure what that is yet, but I know which family it belongs to, so I have a starting point".

Just so you know, we've established that knowing all the plants is impossible, knowing all the genera is unlikely, and know a few of the key families is realistic. Well, what about the next division up from families? Order is the next division up, and there are less orders than families, however the orders have a lot

6

less characteristics in common and wouldn't really be very useful.

The diagram below shows the simplified structure of the classification of all living things:

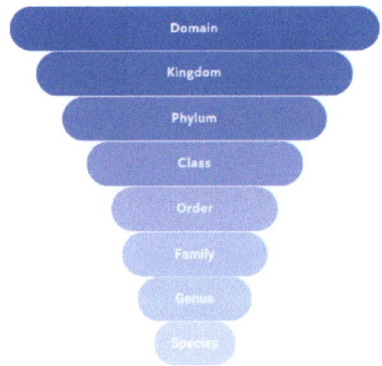

If you want to really get into taxonomy, we could add sub-species, tribes, sub-families, and clades too, but let's not!

Learning Tip: Think about the plants local to you, which you can already identify. If you don't already know, look up the scientific name for them. Now see if you can find the common features of the genus that plant belongs to. You've just turned knowledge of one plant into knowledge of many related plants.

Chapter 2 – Plant families

Plant families are one of the best learning tools and an essential key to accelerate our identification abilities. Whether you're the type of person who learns best by reading, watching, or doing, we all learn best through patterns, and the patterns of plant families are no different. Those patterns can also help you to learn through their exceptions too. For example, nightshade family (*Solanaceae*) plants are generally considered toxic, but two of their food plants potato and tomato produce very similar looking fruit. As we know, tomato fruit is a delicious food crop, whereas the very similar looking potato fruit is very toxic.

Woody nightshade (solanum dulcamara) fruit

I recommend starting with Mint (*Lamiaceae*), Mustard (*Brassicaceae*), Sunflower (*Asteraceae*), Rose (*Rosaceae*), and Parsley (*Apiaceae*). The reference section of this book contains some information on those families.

WHAT IF YOU'RE NOT SURE OF THE FAMILY?

When you're studying a plant you've found and it doesn't seem to fit the families that you know, then you might simply have found a plant outside the families which you now know – so maybe it's time to learn a new family? The next step is to consider using a plant key. Now, before you throw this book in the bin in disgust, I have put together a section about plant keys and how to use them. In that section I'm aiming to demystify keys and make them a little less scary. Finally, if all else fails, and as unlikely as this may seem, it's not impossible that you may have found something new! It does happen sometimes.

OTHER RESOURCES

Mobile apps are the current hot topic, and whilst they have their uses, I am very wary of using them for foraging purposes. Whilst they are accurate sometimes, it's when they get it wrong that concerns me, particularly with families such as the Apiaceae which have some of the best edibles, but also some of the most deadly poisonous too. No app (as of the time of writing) can reliably tell the difference between edible Cow parsley (*Anthriscus sylvestris*) and deadly Hemlock (*Conium maculatum*).

Google images (or any other search engine) relies on the images having been tagged correctly by the author, and as I'm sure you're aware anyone can post just about anything they like on the internet. I have seen the seeds of dangerous giant hogweed (*Heracleum mantegazzianum*) tagged as edible seeds of common hogweed (*Heracleum sphondylium*).

Finally, online groups such as the ones on Facebook. There are some very knowledgeable people on those groups, but then there are some well-meaning, not so knowledgeable people too. Just don't trust any help you're given unless you're confident about the person giving it.

Having said all that's negative about these resources, they can provide a place to start your own research. They might give you an indication of which family to start looking in. Just don't rely on them for things you're planning to eat.

CHAPTER 3 – WHEN NATURE BREAKS THE SYSTEM

No system is infallible, especially when you're trying to impose a human system upon the natural world. Broadly speaking there are two potential issues that this system cannot cover:

1. Humans made this system, not nature, and nature doesn't do as it's told. Trees, plants, and mushrooms don't know our rules, and I suspect if they did, they probably wouldn't care. I'm always finding things growing in a habitat which the guides say they don't grow in.
2. We're learning more and more all the time, so the things we thought we knew can change over time. Species are sometimes moved to other genera, or families, and sometimes entire genera are moved. Nowadays this happens most often with DNA research as we find that plants we assumed are related, are in fact not. Also, the entire species name can change sometimes, and the taxonomists have been known to change them back again.

So having explained how confusing and unreliable the whole system can be, it's still the best system there is, and if you bear all of that in mind when doing your research, you shouldn't run into too many issues. Besides, you can always post it online – there will always be someone keen to point out why you're wrong!

PART 2 – ALL ABOUT PLANTS

When it comes to identifying a plant, reference books are full of botanical jargon, and it can seem a bit intimidating at first. I'm sorry to say that there are some parts that you just have to learn, but they're easy to pick up, and some parts that make sense when you see what they mean.

The following sections explain about

- Plant lifecycles – if you know how and when a plant grows, you can be there to harvest them at the best time,
- and a whole load of terminology relating to leaves, flowers, and fruits, which you may come across when trying to identify plants and trees.

There's quite a lot in there, and it may seem a little overwhelming as you read through, but as you start using this as a reference and only looking up what you need, it'll soon start to sink in.

CHAPTER 4 – PLANT LIFECYCLES

Plants can have different types of lifecycles; they don't all just grow, produce fruit/seeds, die back and then do it all again next year. The most common lifecycles you'll come across are annual, biennial, and perennial.

ANNUALS

Annual plants live their entire lifecycle in one year. The seed germinates, the plant grows, produces fruit and/or seeds, then dies. It is not uncommon to confuse these for perennials, as they often drop their seeds close to where they grow, and the new plants grow in almost the same place.

Chickweed (*Stellaria media*) is a common annual wild food plant which is often found growing in the same place every year, but each year it is a different plant.

Chickweed (Stellaria media) in flower

Biennial plants live their lifecycle over two years. In most cases, the seed germinates, and the plant grows large leaves. Over the first year the plant collects energy from the sun and stores it in its root system. In year two, the plant continues to collect energy until it is ready, then it uses the stored energy to produce flower stems, fruit and/or seed to spread.

Greater burdock (*Arctium lappa*) is a common biennial wild food, prized for its flavoursome root. Being biennial, the best time to collect the root is early in the second year before the flower shoot has begun to grow. In the second year as the flowers die, it produces very sticky, hooked seeds which attach to people and animals to spread far and wide.

Banks of second year greater burdock (Arctium lappa)

Perennial plants live for more than two years, mostly living an annual cycle of seed production (there are exceptions to this). Although trees and woody shrubs are perennial, the term perennial is generally used to refer to non-woody plants.

Yarrow (*Achillea millefolium*) is a common annual wild food plant which comes back year after year with foliage used as a herb and flowers for herbal medicine.

Yarrow (Achillea millefolium) leaves

CHAPTER 5 – LEAF AND LEAF RELATED TERMS

The leaves of a plant allow the plant to absorb moisture, sunlight, and sometimes airborne nutrients (including Carbon Dioxide); They can also allow the plant to release Oxygen as a waste product.

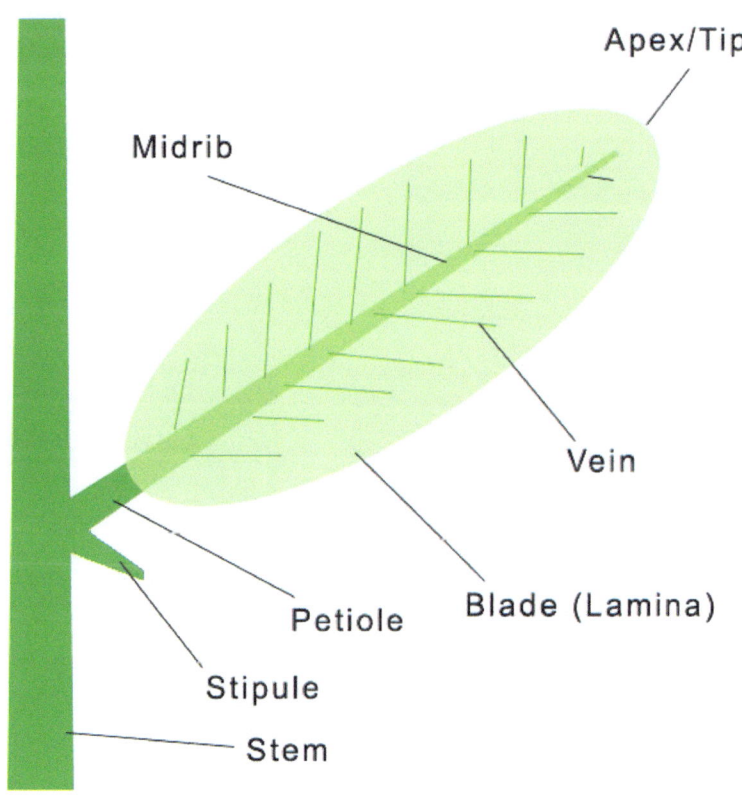

Simple leaf diagram

Let's start with some of the terms that are likely to mean nothing when they're part of text:

Petiole: The petiole is the leaf stem. Sometimes referred to as the leaf stem.

Blade: The bit of the leaf that is left if you remove the petiole.

Leaf petiole and blade of flowering currant (Ribes sanguineum).

Stipules: Stipules are small, leaf-like growths that come out of the base of the petiole. Commonly seen on roses, for example.

Stipules on a wild rose (Rosa canina)

Bracts: Bracts are a modified type of leaf that is usually associated with the flowers of the plant. Bracts are really obvious on Linden trees when they are in flower because they are a much paler green, usually in very large numbers and attached to the flower stems.

Linden flower (Tilia × europaea) showing the attached bract

Node: The node is where a petiole is attached to the plant stem.

Internode: The space between nodes.

Midrib: The leaf may have a single prominent "vein" running from base to tip. This is known as the midrib.

Axil: The Axil is part of the node, specifically the upper angle where the petiole joins the stem.

Seed leaves: When a seed first germinates and emerges, if often has one or two seed leaves which typically serve to give the seedling a head-start before its true leaves can develop. Sometimes also called Cotyledons.

Cotyledons (seed leaves) of hornbeam tree (Carpinus betulus)

LEAF DESCRIPTIONS

At its simplest, a leaf is either simple or compound. Simple meaning a stereotypical leaf shape, compound meaning many parts, or leaflets, joined to a single stem. Let's take a look at some common simple leaf shapes:

SIMPLE LEAF SHAPES

CORDATE

Heart-shaped, stem in indent of heart.

Example: Common lime (*Tilia x europaea*)

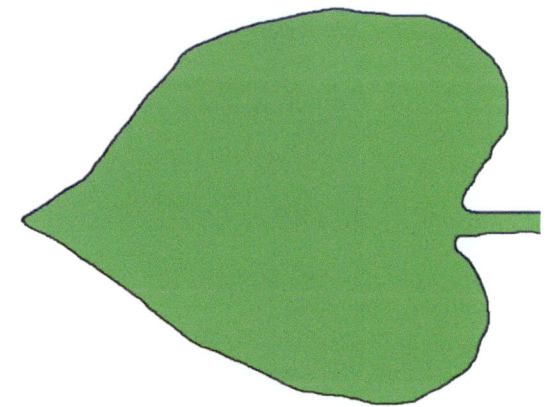 

ROUND (ORBICULAR)

Example: Aspen (*Populus tremula*)

Oval (ovate)

Egg-shaped with the wide end near the petiole. Obovate is also egg-shaped but with the narrow end near the petiole.

Example: Ground Elder (*Aegopodium podagraria*)

Lanceolate (lance-like)

Pointed at both ends.

Example: Ribwort plantain (*Plantago lanceolata*)

Mostly parallel sides with pointed ends.

Example: Pendulous sedge (*Carex pendula*)

LEAF EDGE (OR MARGIN) FORMS

ENTIRE

Even and smooth.

Example: English ivy (*Hedera helix*)

CRENATE

Like the rounded crenelations on the top of a castle wall.

Example: Stinging nettle (*Urtica dioica*)

Dentate (like teeth)

Like symmetrical teeth.

Example: Wild strawberry (*Fragaria vesca*)

Serrate (like a serrated knife)

Forward-pointing teeth. Double-serrate has smaller teeth in between the main teeth.

Example: Wild rose (*Rosa canina*)

Indented, but not as far as the centreline.

Example: Common hogweed (*Heracleum sphondylium*)

LEAF ARRANGEMENT ON THE STEM

BASAL ROSETTE

(leaves spreading out from the base of the stem on the ground. – Note: the shape of the basal leaves are often a totally different shape to those further up the stem).
Example: Dandelion (*Taraxacum officinale*)

OPPOSITE

The leaves grow on opposite sides of the stem from the same position, usually in pairs.

Example: Water mint (*Mentha aquatica*)

Leaves grow on opposite sides of the stem, offset from each other.

Example: Crab apple (*Malus sylvestris*)

WHORLED

Leaves grow around the stem from the same position.

Example: Self-Heal (*Prunella vulgaris*)

COMPOUND LEAF ARRANGEMENT

BIFOLIATE

Two leaflets forming one leaf.

TRIFOLIATE

Three leaflets growing forming one leaf.

Example: Wood sorrel (*Oxalis acetosella*).

More than three leaflets (usually five) radiating out from the same spot, forming one leaf.

Example: Motherwort (*Leonurus cardiaca*)

EVEN PINNATE

Two or more pairs of leaflets growing opposite, each pair from the same position on the stem.

Example: Mastic tree (*Pistacia lentiscus*)

Two or more pairs of leaflets growing opposite, each pair from the same position on the stem, and a terminal leaflet at the end making the total number of leaflets in this leaf an odd number.

Example: Rowan tree (*Sorbus aucuparia*)

BIPINNATE

A leaf which divides into two or more stems, each containing a pinnate arrangement of leaflets.

Example: Wild angelica (*Angelica sylvestris*)

A leaf which divides into two or more stems, each dividing into two or more stems again, each containing a pinnate arrangement of leaflets.

Example: Yarrow (*Achillea millefolium*)

CHAPTER 6 – FLOWERS AND FLOWER RELATED TERMS

Flowers are the primary method of reproduction in plants of the modern world, allowing the plant to produce seeds and sometimes fruit, which in turn allows it to spread and create new plants.

Flowers can be bisexual, having both male and female parts in one flower; Or, where plants have separate male and female flowers, they can both appear on one plant (**monoecious**) or on separate plants (**dioecious**).

Knowing this can be useful when looking for fruit, for example. Male flowers do not produce fruit so there's no point returning to a male plant waiting for fruit or seeds. Stinging nettles (*Urticaria dioica*) has both male and female plants, but the treasured, nutritious stinging nettle seeds can only be gathered from female plants.

Male stinging nettle flowers

PARTS OF A FLOWER

Why care about which bit is the pistil and which bit is the anther? When you get onto using keys and you need to know how many pistils, and what colour are the anthers it really does help to know what the heck it's talking about.

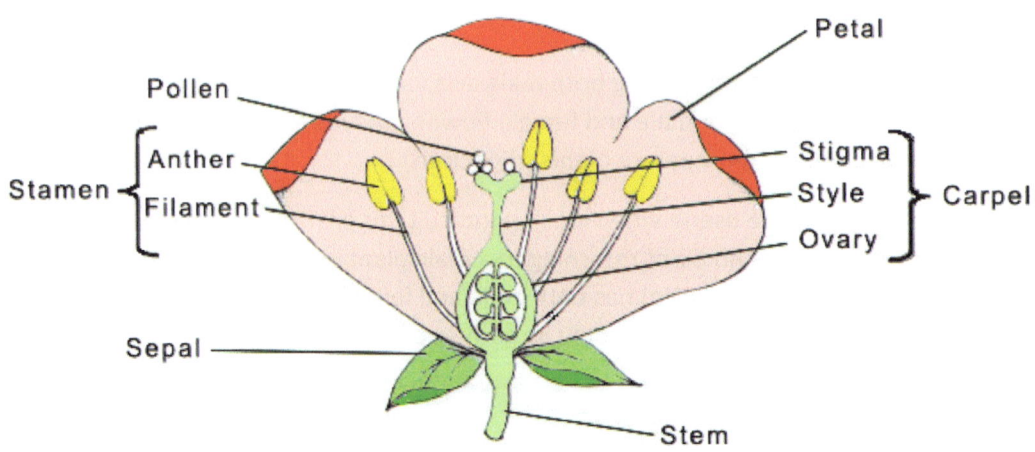

Parts of a bisexual flower

Sepal: Imagine an unopened, green flower bud. When the flower opens and those green parts separate, those are the sepals. Sometimes they fall off when the flower opens, sometimes they remain.

Calyx: Calyx is the term used to refer to all of the sepals.

Petal: The petals are the usually showy, colourful parts of the flower which open up.

Corolla: Corolla is the term used to refer to all of the petals.

Stamen: The stamen are the male parts of the flower including the filament and the anther.

Anther: The anther is the pollen producing part, which can be different colours.

Filament: The part of a flower which supports the anther.

Carpel: The carpel is the female part of the flower comprising the ovary, style, and stigma.

Stigma: The female part of a flower which receives the pollen.

Style: The stem of the female part which delivers the pollen to the ovary.

Ovary: The female part of the flower which produces ovules.

Pistil: You may occasionally hear mention of the pistil. This comprises the stigma and style, the parts responsible for producing an ovule (egg) in the ovary.

Stem: Also known as the peduncle when referring to flowers and fruit.

Three-Cornered Leek (*Allium triquetrum*) flower.

Inflorescences

Inflorescence is the term used to describe how a flower(s) is arranged. This can help when you start to see that an inflorescence can be a single flower, or many flowers clustered together in varying arrangements.

This subject can get very complex with many different arrangements, sub-arrangements, and compound arrangements, but I've tried to keep it to the simplest and most common types in this book.

RACEME

Unbranched inflorescence with flowers on short stems along its main stem. For example, Rosebay Willowherb (*Chamaenerion angustifolium*).

SPIKE

A type of raceme whose flowers have no stems. For example, Ribwort Plantain (*Plantago lanceolata*).

CORYMB

A type of raceme whose outer stems are longer than its inner stems, giving it a flat or convex top. For example, Elder (*Sambucus nigra*).

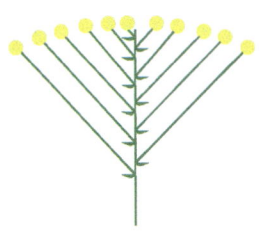 

UMBEL

A type of raceme with a short stem. The multiple flower stems all originate from the end of the main stem and are of similar length. Typical of *Apiaceae* (formerly *Umbelliferae*) family plants for example, Hemlock Water-Dropwort (*Oenanthe crocata*).

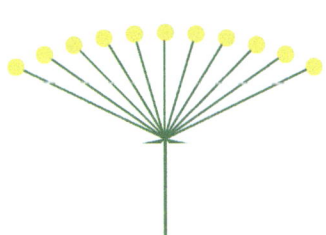 

SPADIX

Similar to a spike but the tightly clustered flowers are arranged around a specialised, fleshy bract called a spathe. Typical of *Araceae* family plants for example Lords and Ladies (*Arum maculatum*).

HEAD

Also known as a capitulum is a very compressed raceme where single, stemless flowers are on an enlarged stem. Typical of *Dipsacaceae* family plants (teasels). A further compressed and flattened form of capitulum is typical of the *Asteraceae* family for example, Chamomile (*Matricaria chamomilla*).

CATKIN

A scaly drooping spike or raceme. Typical on many trees in the *Betulaceae* family for example, Alder (*Alnus glutinosa*).

CYME

A cyme is a flat-topped inflorescence where the central flowers open first, followed by the outer flowers for example Red Campion (*Silene dioica*). Cymes can be further divided into compound arrangements which are beyond the scope of this book.

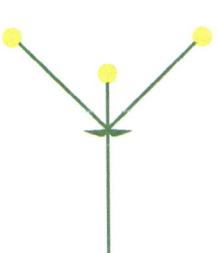

A panicle is a compound raceme which is many branched and has only one flower per stem for example, Pendulous Sedge (*Carex pendula*). A similar arrangement with multiple flowers per flower stem is a compound spike.

CHAPTER 7 – FRUIT AND FRUIT RELATED TERMS

Botanically speaking, a fruit is the seed-bearing ripened ovary of a flowering plant which may or may not also include other parts of the plant. The purpose of a fruit is to protect and/or distribute the seeds. So again, botanically speaking any food that has seeds in it is a fruit, including tomatoes, cucumber, nuts, and so on. Grocery speaking, you can call it a vegetable if that's what you're used to, but it's still a fruit.

Fruit can be divided into fleshy or dry fruit. Fleshy would be apples, tomatoes, berries, etc. Dry would be dandelion fluff, nuts, poppy heads, etc.

Dry fruit can be further divided by how they disseminate their seeds, but we don't really need to go that far. Fruit can also be classified based on how it develops from the plants and which bits of the flower(s) are incorporated etc. but that's a level of classification we don't really need. In the descriptions below I've tried to stick to the useful terms, and the ones you may need to understand from reference books.

Cherry plums (Prunus cerasifera)

"Simple" means that the fruit developed from a single flower with a single pistil, and includes:

Achene – e.g. fat-hen (*Chenopodium album)*, strawberry (*Fragaria vesca*) I know! See accessory fruit below for an explanation.

Fat hen (Chenopodium album) achenes

An Achene is a dry fruit which does not open to release its single seed.

Capsule - e.g. poppy (*Papaver rhoeas*), Brazil nut (*Bertholletia excelsa*)

A Capsule is a dry fruit which opens up to release more than one seed.

Poppy (Papaver rhoeas) capsule

Cypsela - e.g. dandelion (*Taraxacum officinale*)

A Cypsela is a dry fruit which does not open to release its single seed, characteristic of the Asteraceae family.

Dandelion (Taraxacum officinale) cypsela

Follicle – e.g. magnolia (*Magnolia* spp.)

A Follicle is a dry, single compartment containing two or more seeds.

Magnolia (Magnolia spp) follicles

Legume – e.g. clover (*Trifolium repens*)

A legume is the fruit of a plant in the *Fabaceae* family.

Clover (Trifolium repens) legumes

Nut – e.g. hazelnut (*Corylus avellana*)

A nut is a fruit consisting of a hard shell which protects a kernel inside, which is usually edible.

Hazel (Corylus avellana) nuts

Samara – e.g. ash (*Fraxinus excelsior*)

A samara is a winged achene, with a flattened wing of papery tissue which allows it to be carried further away on the wind.

Hornbeam (Carpinus betulus) samara

Schizocarp – e.g. carrot (*Daucus carota*)

A schizocarp is a dry fruit which when mature, splits up into multiple mericarps.

Wild carrot (Daucus carota) schizocarps

Silique – e.g. hairy bittercress (*Cardamine hirsute*)

A silique is a type of fruit (seed capsule) having two fused carpels with the length being more than three times the width.

Hairy bittercress (Cardamine hirsute) silique

Silicle – e.g. shepherd's purse (*Capsella bursa-pastoris*)

A silicle is a type of fruit (seed capsule) having two fused carpels with the length being less than three times the width.

Shepherd's purse (Capsella bursa-pastoris) silicle

Utricle – e.g. sheep's sorrel (*Rumex acetosella*)

A utricle is a one-seeded fruit that has developed from one flower having a single ovary. The ovary wall becomes bladdery or inflated at maturity.

Sheep's sorrel (Rumex acetosella) utricles

SIMPLE FLESHY FRUIT

Simple fleshy fruits include true berries, drupes, and pomes.

Berry – The most common type of edible fruit formed from a single ovary, and a frequently confused term. For example, grapes, cucumbers, and bananas are berries, whereas strawberries, and raspberries are not!

e.g. gooseberry (*Ribes uva-crispa*)

Wild gooseberries (Ribes uva-crispa)

Drupe – Also known as stone-fruit, has a hardened wall surrounding the seed.

e.g. sloe (*Prunus spinosa*)

Ripe sloes (Prunus spinosa)

Pome – Has a core containing the seeds, surrounded by a tough membrane, then the fleshy, edible layer.

e.g. crab apples (*Malus sylvestris*)

Windfall crab apples (Malus sylvestris)

AGGREGATE FRUIT

An aggregate fruit develops from a single flower with multiple pistils. As each "fruitlet" develops from its pistil, they join together to form the aggregate fruit. Different plant species can be an aggregation of different types of fruit, for example Raspberries (*Rubus idaeus*) are an aggregation of drupelets, and *Magnolia* (Magnolia spp.) fruit are an aggregation of follicles.

Wild raspberry (Rubus idaeus)

MULTIPLE FRUIT

A multiple fruit develops from multiple, closely clustered flowers where the fruitlets merge together as they develop. For example, pineapple (*Ananas comosus*), fig (*Ficus carica*), and mulberry (*Morus nigra*) are all multiple fruits.

Black mulberry (Morus nigra)

An accessory fruit is one where tissues from other parts of the flower are incorporated into the fruit, such as petals and sepals. Accessory fruits can occur as simple, aggregate, or multiple fruits, so you can have a simple-accessory fruit such as an apple (*Malus sylvestris*), a multiple-accessory fruit such as pineapple (*Ananas comusus*), or an aggregate-accessory fruit such as wild strawberry (*Fragaria vesca*).

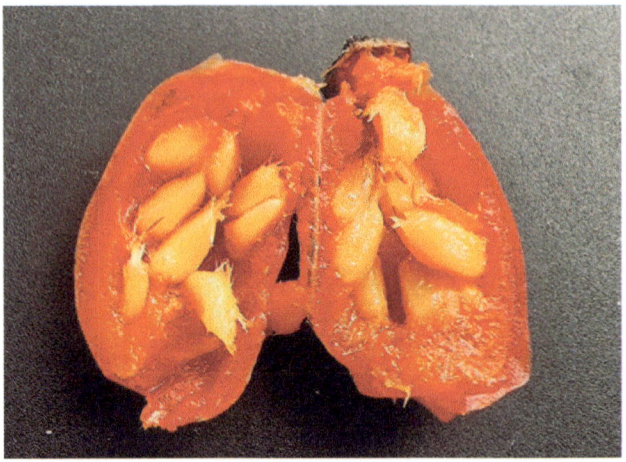

Rosehip (Rosa canina)

PART 3 – PLANT KEYS

Some reference books have no key, instead they group their trees, plants, or fungi by family, and you can look in the family until you find the one you're looking for – hence this family identification system. Some however, will have keys which can take you from nothing to species by asking a series of questions to narrow down the possibilities. For example, if you were looking up a tree the first question in the key might be: "Does it have needle-like leaves? Yes, go to section 3. No, go to question 2".

Finding a good key can be like hitting the fast-forward button on your identification task; but sometimes an identification key can be an impenetrable barrier of tiny text, botanical terminology, and frankly just really difficult to even find the starting point. I won't say that there are good and bad keys, but some keys are definitely easier to use than others.

Flower-based keys are the most common, but you'll also find fruit-based keys, tree keys, fungus keys, and vegetation-based keys for when there are no flowers or fruit to see.

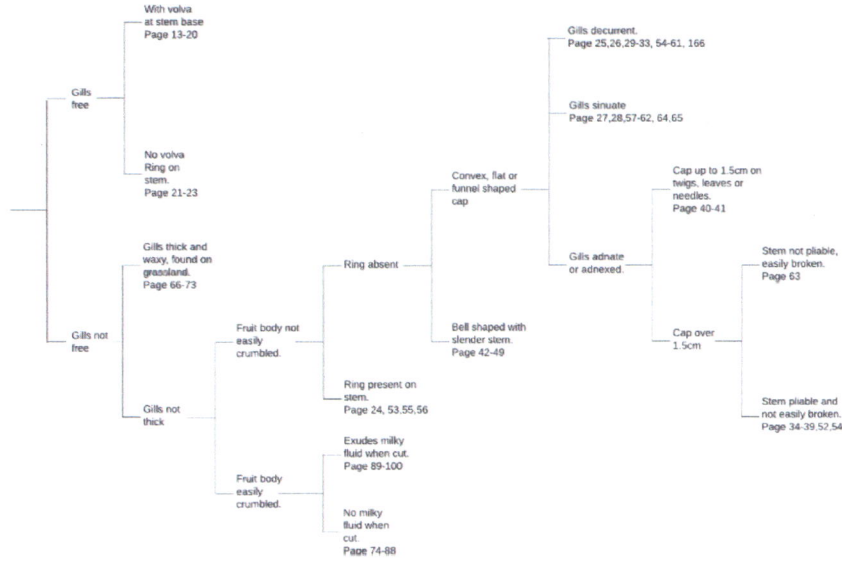

Simplified key example

CHAPTER 8 – HOW TO USE PLANT KEYS

Plant keys, usually from books though sometimes found on websites, almost always come with their own instructions because there is no single way in which all keys are arranged. This is a good thing, because over time you can find a preferred key to refer to, and bad because if you do need to use another key you'll need to work out how it operates first.

However, there is one practice that will help you with all keys, and I encourage you to get in the habit of doing it as it will also help you build a connection, understanding, and solid memory of the plant in question.

PROFILE YOUR PLANT

The standard keys are flower keys, but this practice applies equally to flowers, leaves, fruit, and fungi.

Begin with a notepad, or your preferred digital device and record the plant profile:

1. Describe the larger plant that the part has been taken from.
2. Describe the environment that it was growing in.
3. Note any nearby trees and plants.
4. Identify all of the relevant parts for the key (all the flower parts for example).
5. Record each part, along with the colour, number of, and arrangement of its parts.
   a. You can deconstruct the plant as you do this, but always start from the outside and work in.
6. Note any absences and if possible, ignore any missing parts that may have fallen off or been damaged.

Check the back of this book for a handy plant description template to help you remember everything that you need to write down.

Now you have a detailed profile of your plant/plant parts, you can begin to look at the key. Always start at the beginning and don't be tempted to jump ahead.

My top tip for using keys, is to remember that they don't always give the desired result, so make sure not to try to make the key fit the plant. Sometimes the plant won't fit any of the options given; in that case I usually pick the most likely option and if it doesn't work out, I go back to that decision point and try again.

PART 4 – REFERENCE GUIDE

Having rather gone on about keys and how to use them, I present a reference guide to some of the common and widespread families which you might come across, however, this is not in any way a key. There is a simple reason for that, which is that a key should be extensive and should aim to always be able to deliver an answer to the question of identification. Therefore, the creation of a botanical key is a mammoth undertaking and well beyond the scope of this book.

The reference guide will introduce five of the most common and widespread plant families: Lamiaceae, Brassicaceae, Asteraceae, Rosaceae, and Apiaceae, and for each give a few examples of common, edible plants.

Wild strawberries (Fragaria vesca)

CHAPTER 9 - *LAMIACEAE*

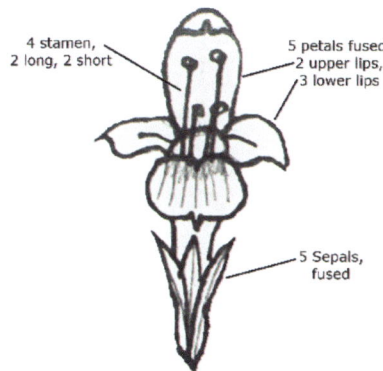

4 stamen,
2 long, 2 short

5 petals fused,
2 upper lips,
3 lower lips

5 Sepals,
fused

Black horehound (Ballota nigra) line drawing

The *Lamiaceae* (formerly known as *Labiatae*), or mint, or sage, or deadnettle family is very common, widespread and has some very recognisable features.

If your candidate has a stalk with a square cross-section, with simple and opposite leaves, you may have a member of the mint family. The flowers are distinctive, if a little difficult to describe. They have five sepals, joined together to form a tube, and five petals also joined together in the sepal tube, spreading out to become two fused lobes at the top, and three fused lobes at the bottom (often referred to as upper and lower lips). Further to this, they also have two pairs of stamens, with one pair being longer than the other.

Mint family plants are usually, but not always, aromatic too. Crush a few leaves and they give up their smell as the volatile oils are released. They are often hairy too, as you can see from the picture of self-heal below.

Lamiaceae contain about 236 genera and around 7500 species.

Self-heal (Prunella vulgaris)

A feature of the leaves of the *Lamiaceae* family is that as well as being in opposite pairs, each pair grows at 90 degrees to the previous pair.

The family has over 7000 species in over 200 genera, including many culinary herbs such as basil (*Ocimum*), mint (*Mentha*), rosemary (*Rosmarinus*), sage (*Salvia*), oregano (*Origanum*), and thyme (*Thymus*).

GLECHOMA

Eight species, perennial herbs, leaves on long petioles, two or more blue-violet flowers growing from leaf axils.

Ground ivy, *G. hederacea*. Round to kidney-shaped leaves with crenate edges. Mauve to purple flowers with darker spots on the lower lip. Leaves are usually present year-round, flowers from March to June and again in the Autumn. Usually grows along the ground but can grow upright too. Has a distinctive smell and a pungent taste, quite a "meaty" flavour with a very weak background hint of mint.

Ground ivy (Glechoma hederacea)

15 species of hemp-nettles, annual herbs, upright, ovate leaves with pointed tips, generally considered poisonous.

Common hemp-nettle, *G. tetrahit*. Ovate leaves with an elongated tip and crenate edges. Pinkish flowers with darker spots on the lower lip. Usually in flower from March to September. Grows upright, up to 70cm tall. Has an unpleasant, musty smell. Not much is known about its edibility; A French publication says it is, whilst a Russian publication say that it is poisonous and causes paralysis. Not worth the risk until more research is available.

Hemp nettle (Galeopsis tetrahit)

HYSSOPUS

Seven species of hyssops, herbaceous/semi-woody, upright, narrow oblong leaves with an entire edge. Small blue flowers growing from leaf axils in summer. Aromatic and often used in food and herbal medicine.

Hyssop, *H. officinalis*. Lanceolate leaves with entire edges. Blue fragrant flowers in the summer. Grows upright, up to 60cm tall. Has an intense minty smell and is used in the production of the liqueur Chartreuse, and in some versions of Absinthe.

Hyssop (Hyssopus officinalis)

30 or more species of deadnettles, annual and perennial herbs of varying growth habits and flower colours. The name probably derives from the similarity of *L. album* to the unrelated stinging nettles (*Urtica dioica*), but without stings makes them "dead".

White deadnettle, *L. album*. Herbaceous perennial, grows upright up to 100cm tall. Leaves triangular with a rounded base and serrated edge. Petiole up to 5cm long. White flowers grow around the leaf axils. Can commonly be found growing with stinging nettles. Completely edible, but not much smell or flavour.

White deadnettle (Lamium album)

Yellow archangel, *L. galeobdolon*. Sometimes known as yellow deadnettle. Herbaceous perennial, grows upright up to 80cm tall. The leaves are roughly ovate with a pointed tip and crenate edges, and a short petiole. The flowers are yellow and grow around the leaf axils. Completely edible, but much like the similar White deadnettle, very little smell or flavour.

Yellow archangel (Lamium galeobdolon)

Red deadnettle, *L. purpureum*. Herbaceous annual, grows upright up to 20cm tall. Unlike its namesakes, white and yellow deadnettle, red deadnettle leaves do not look like stinging nettles. Leaves are stalked, kidney shaped and crenated, and the upper leaves tend to take on a purplish colour. Pink-purple flowers from early spring, but can flower all year, including in mild winters. Completely edible, not much smell, pleasant slightly bitter flavour.

Red deadnettle (Lamium purpureum)

24 species of motherworts, annual and perennial herbs with pale pink to purplish-red flowers. Depending on where they are found, the species are used in traditional Chinese medicine and western herbal medicine.

Motherwort, *L. cardiaca*. Herbaceous perennial, grows upright up to 1m tall. A square, hairy stem, often purple around the leaf nodes. Opposite, palmately lobed leaves with serrated edges and a long petiole. Green and slightly hairy on top, smooth and grey underneath. The flowers grow in the leaf axils and have three lobed bracts. Pink to lilac flowers from mid to late summer. Edible and nutritious, it has also been used for flavouring beer.

Motherwort (Leonurus cardiaca)

49 species of horehounds, annual and perennial herbs, generally native to the Mediterranean region.

White horehound, *M. vulgare*. Herbaceous perennial grows up to 45cm tall. Grey-green leaves with deep "crinkles" and covered in soft hairs. The flowers are white and in clusters around the top of the main stem. Another herb used for beer flavouring, horehound is also the main ingredient in horehound candy drops which are bitter and sweet.

White horehound (Marrubium vulgare)

24 species of mints, almost always perennial and aromatic with almost universal distribution. Used widely in food and medicine.

Water mint, *M. aquatica*. Herbaceous perennial grows up to 90cm tall and has wide spreading rhizomes. The stems are square cross-section and can be either green or purple or a mixture. The leaves are wide, green (sometimes purplish), opposite and toothed. The flowers are tiny and densely packed on a terminal inflorescence. The flowers vary from pink to lilac to purple, and every part of the plant has a distinctly minty smell.

Water mint (Mentha aquatica)

13 species of herbaceous plants, mostly found in the northern hemisphere. Edible, and prized for their use in herbal medicine hence known as the self-heals or allheals.

Self-heal, *P. vulgaris*. Herbaceous perennial grows up to 30cm high but is usually much smaller. It has tough, reddish stems, with a square cross-section. The small, purple and white flowers grow from a club-like terminal inflorescence in a whorled cluster. It mostly flowers throughout the summer.

Self-help (Prunella vulgaris)

The largest genus of plants in the family Lamiaceae, with nearly 1000 species of shrubs, herbaceous perennials, and annuals. Found almost everywhere but the driest and coldest regions and includes the kitchen herbs sage and rosemary.

Sage, *S. officinalis*. Perennial evergreen shrub with woody stems, grows up to 60cm tall and wide. It has grey-green oblong leaves which are nearly white underneath and covered with soft hairs. Lavendar coloured flowers are most common, although they can be white, pink, or purple. It flowers in late spring and summer. Native to the Mediterranean region but spread almost everywhere due to its culinary uses.

Common sage (Salvia officinalis)

365 species of hedgenettles, woundworts, and betony, found across most of the world. The species are all shrubs and annual or perennial herbs and are valued for their medicinal uses.

Field woundwort, *S. arvensis*. Annual herbaceous which grows up to 45cm long either creeping or upright. Pale green/reddish, angular stems with many small hairs. Leaves are in opposite pairs, up to 4 cm long, oval and have toothed edges, and are hairy on both sides. The terminal inflorescence has two opposite clusters of up to 7 white, pink, or purple flowers. It has a kind of musty smell and flavour with a hint of mint and is better known for its medicinal qualities.

Field woundwort (Stachys arvensis)

CHAPTER 10 - *BRASSICACEAE*

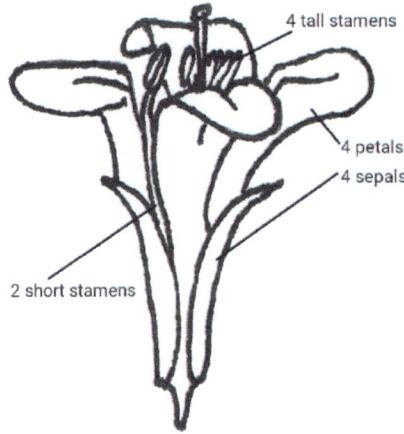

Typical mustard family flower line drawing

The Brassicaceae family also known as the mustards, the crucifers, or the cabbage family. Most are herbaceous plants, while some are shrubs. The leaves are simple (although are sometimes deeply incised), lack stipules, and appear alternately on stems or in rosettes. The inflorescences are terminal and lack bracts. The flowers have four free sepals, four free alternating petals, two shorter free stamens and four longer free stamens. The fruit has seeds in rows, divided by a thin wall (or septum).

The family contains 372 genera and 4,060 accepted species.

ALLIARIA

2 species of garlic mustard native to temperate Eurasia.

Garlic mustard, *A. petiolata.* herbaceous biennial plant growing from a deeply growing, thin, white taproot that is scented like horseradish. In the first year, plants appear as a rosette of green leaves close to the ground; these rosettes remain green through the winter and develop into mature flowering plants the following spring. Second year plants grow from 30–100 cm tall and rarely to 130 cm. The young leaves are good chopped into cooked food such as pasta, or mashed potatoes. The green seeds are a delicious sweet mustard treat and the dark brown mature seeds make a great pungent mustard substitute.

Garlic mustard (Alliaria petiolata)

3 species of horseradish native to the east northern hemisphere.

Horseradish, *A. rusticana.* Perennial plant growing up to 1.5m tall. Hairless bright green unlobed leaves up to 1m long, with wavy edges. The white four-petalled flowers are scented and are borne in dense panicles. Established plants may form extensive patches. Whilst cultivated for its spicy, pungent root, the young leaves are an often-overlooked delicate mustard addition to salads.

Horseradish (Armoracia rusticana)

7 species of Shepherd's purse native to temperate Eurasia. The name is said to derive from Latin capsa, a box or case, alluding to fruit resembling a medieval wallet or purse.

Shepherd's purse, *C. bursa-pastoris.* Annual herbaceous plant growing up to 0.4m tall. It is described as the second-most prolific wild plant in the world and is common on cultivated ground and waysides and meadows. It grows from a basal rosette of lobed leaves with a single stem which has a few pointed leaves that partially grasp the stem. The flowers, which appear all year round, are white and small, 2.5 mm, with four petals and six stamens. They are in loose racemes, and produce flattened, two-chambered seed pods known as silicles, which are triangular to heart-shaped, each containing several seeds.

Shepherd's purse (Capsella bursa-pastoris)

263 species of bittercresses and toothworts. All annual and perennial herbs found in most environments apart from the Antarctic. The name Cardamine is derived from the Greek kardamine, watercress, from kardamon, pepper grass.

Hairy bittercress, *C. hirsuta.* Strictly classed as an annual, hairy bittercress can complete multiple generations in one year. It grows a rosette of leaves at the base of the stem, while there may be few leaves on the upright stem. The leaves in the rosette are pinnately divided into 8–15 leaflets which have short stems connecting them to the petiole. These basal leaves are often 3.5–15 cm long. The leaflets are round to ovate and may have smooth or dentate edges. The leaflet at the tip of the leaf (terminal leaflet) is larger than the other leaflets and round to kidney shaped. The small white flowers are borne in a raceme without any bracts, soon followed by the seeds. The flowers have 4 white petals which are 1.5–4.5 mm long and spatulate shaped. The flowers also have 4 stamens of equal height instead of the 6 which are found in most closely related plants.

Hairy bittercress (Cardamine hirsuta)

35 species of annual and perennial plants native to Europe, Asia and eastern Africa. They have dense racemes of tiny white or yellow flowers on (mostly leafless) stems above the basal leaves. The word "crambe" derives, via the Latin crambe, from the Greek, a kind of cabbage.

Sea kale, *C. maritima.* Perennial, mound-forming, salt-tolerant plant growing up to 75cm tall. It has large fleshy blue-green leaves and abundant white flowers. The globular pods contain a single seed. The plant is related to the cabbage and was first cultivated as a vegetable in Britain around the turn of the 18th century. The blanched stems are eaten as a vegetable.

Sea kale (Crambe maritima)

4 species of perennial watercresses or yellowcresses. Native to Europe through central Asia, Africa, and North America. Usually with yellow flowers and a peppery flavour.

Watercress, *N. officinale*. Perennial, fast-growing plant up to 1m creeping. The hollow stems of watercress float in water. The leaf structure is pinnately compound. Small, white, and green inflorescences are produced in clusters. It is one of the oldest known leaf vegetables consumed by humans. Watercress crops grown in the presence of manure can be an environment for parasites such as the liver fluke which can be very dangerous if ingested, which is why some people avoid foraging for it.

Watercress (Nasturtium officinale)

4 species of yellow flowering wild mustards.

Charlock, *R. arvense.* Annual plant with yellow flowers also known as the charlock mustard, field mustard, or wild mustard, growing to 80cm tall. The stems are erect, and branched, with coarse hairs near the base. The leaves are petiolate 1–4 cm long. The basal leaves are oblong, oval, lanceolate, 4–18 cm long, 2–5 cm wide. The stem leaves are much smaller and are short petiolate to sessile. It blooms from May to September. The inflorescence is a raceme made up of yellow flowers having four petals. The fruit is a silique 3–5 cm long with a beak 1–2 cm long.

Charlock (Rhamphospermum arvense)

CHAPTER 11 - *ASTERACEAE*

Typical Asteraceae family line drawing

The *Asteraceae* family, also known as the aster, daisy, composite, or sunflower family is the largest family of flowering plants with over 1900 genera and over 32,000 species. Most species of Asteraceae are annual, biennial, or perennial herbaceous plants, but there are also shrubs, vines, and trees. The family has a widespread distribution including every continent except Antarctica.

Their common characteristic is their distinctive flower head, known as a capitulum, consisting of many (sometimes hundreds) of individual florets enclosed by protective bracts.

The family is economically important for food uses, and for decorative purposes, but there are a great many wild examples that can be foraged.

ACHILLEA

135 species of yarrows, native to most of the northern hemisphere and introduced widely in the southern hemisphere. They typically have hairy, frilly, aromatic leaves and large, flat clusters of flowers at the end of stems.

Yarrow, *A. millefolium.* Perennial herb which grows upright with one or more stems, and can grow to up to 0.5m. As well as spreading seeds, it also spreads via rhizomes. The leaves get smaller as they go up the stem and can be from 5-20cm. Leaves are fern-like and divided bipinnately or tripinnately, hence the name "millefolium" or "thousand leaves". The inflorescence contains disc and ray flowers from white to pink from March to October. Other colours may be found in the wild, but they will most likely be garden escapees. There are 3 to 8 ray flowers, which are 3mm long and ovate to round. The tiny disk flowers range from 10 to 40. The inflorescence is produced in a flat-topped capitulum cluster, and the inflorescences are visited by many insects.

Yarrow (Achillea millefolium)

44 Species of Burdocks, native to the Europe and Asia, and widely introduced worldwide. They generally have large, coarse, and ovate leaves up to 70cm long and woolly underneath. Leafstalks are generally hollow and the species mostly flower from July through October. Burdock seeds have hooked burrs which aid seed dispersal.

Burdock, *A. lappa.* Biennial, large herbaceous plant growing up to 3m tall. It has large, alternating, wavy-edged leaves with a long petiole and hairy underside. The flowers are purple and grouped in globe-shaped capitulum. They are surrounded by a series of bracts which curve to form a hook. These bracts allow the seeds to attached to the fur or clothes of passing animals or people. The long taproot is the interesting part for foraging. They can grow up to 1m long and when cooked are crisp and taste sweet and slightly pungent.

Greater burdock (Arctium lappa)

477 species of various plants including Mugwort (*A. vulgaris*), wormwood (*A. absinthium*), and sagebrush (*A. tridentata*). It includes tough herbaceous plants and shrubs which are known for their strong essential oils. The species are very widespread in both hemispheres but prefer dry conditions.

Mugwort, A. *vulgaris*. Perennial herbaceous plant growing up to 2m tall. They have an extensive rhizome system which is their main method of spreading. The leaves are 5-20cm long, dark green, pinnate and have dense, white hairs on the underside making it appear silvery. The stem sometimes has a reddish tinge to it. The 5mm panicle flowerheads are yellow-brown and it flowers from summer to autumn. It has a slightly bitter taste and a very nice smell. The plant is valued for tea and for smoking which apparently induces lucid dreaming. I've used it for flavouring ales and it's one of my favourites.

Common Mugwort (Artemisia vulgaris)

14 species of daisies native to Europe, the Mediterranean, and northern Africa. Widely introduced to north America and the rest of the world. Mostly perennials, they have simple erect stems up to 20cm tall. Most species have basal leaves, and all have one flower head per stem.

Daisy, B. *perennis.* Perennial herbaceous plant growing up to 20cm. Rosettes of spoon-shaped leaves up to 5cm long. The stems are up to 10cm tall and leafless. The flowerheads are composite with white ray florets (sometimes tipped red) and yellow disc florets. The leaves and flowers are edible, young ones are the best. Infused in oil it is sometimes known as English Arnica for its bruise reducing properties.

Daisy (Bellis perennis)

377 species of perennial and biennial plants known as plume thistles. Mostly native to Eurasia and north Africa with some species from North Africa. They are known for their bright flower heads, usually purple, pink, yellow, or white. They have erect stems and prickly leaves, with an enlarged flower base with many prickles.

Spear thistle*, C. vulgare.* Annual or biennial tall plant growing up to 1.5m tall. They have a basal rosette of leaves which grow up to 30cm long. The stem leaves are grey-green, deeply lobed, and sharply spined. The inflorescence is up to 5cm diameter, pink-purple with only one type of floret (i.e. no separate ray and disc florets). The seeds have a feathery pappus which assist in wind dispersal. When you find one with a thick stem, you can strip the spines off and eat it raw like celery. At the base of the inflorescence is a penny-sized, carbohydrate rich section which tastes nutty.

Spear thistle (Cirsium vulgare)

41 species of hawkbits, native to Europe and north Africa, introduced in north America and Australia. Their English name derives from the mediaeval belief that hawks ate the plant to improve their eyesight.

Rough hawkbit, *L. hispidus.* Perennial herb native to Europe and introduced to north America, growing up to 0.5m tall. Forms a rosette of hairy, lobed leaves, with leafless, unbranched stalks bearing yellow, dandelion-like flower heads in summer and early autumn. The young leaves can be eaten raw or cooked, they're not great but they are available through the winter.

Rough hawkbit (Leontodon hispidus)

5 species of mayweeds and chamomiles with widespread temperate distribution. They are hardy, pleasantly aromatic annuals, hairless, with branched stems and they tend to be very leafy.

Pineapple weed, *M. discoidea*. Annual plant native to Northeast Asia, grows up to 40cm long and can grow erect but mostly lies along the ground. The leaves are finely divided pinnately and sweet smelling when crushed. The flower head is cone shaped packed with yellowish green floret, and with no ray florets. The flowers smell strongly of pineapple and chamomile. The young leaves and flowers are used in salads and make a sweet herbal tea. They can also be infused in alcohol and syrups.

Pineapple weed (Matricaria discoidea)

German chamomile, *M. chamomilla.* Perennial herb native to southern and eastern Europe but found on every continent. It has a branched, smooth stem and can grow up to 60cm. The leaves are long, thin and bipinnate. The flowers are made up of yellow disc florets around a swollen, hollow head, and white ray florets. Chamomile is mostly used to make tea, but like pineapple weed the young leaves and flowers can be used in salads.

German chamomile (Matricaria chamomilla)

98 species of sow thistles, native across Eurasia, Africa, and Australia. They are annual, biennial, and perennial herbs. They have soft, lobed leaves beginning in a basal rosette and the stem contains a milky latex. Flower heads are yellow and range in size.

Field sow-thistle, *S. arvensis.* Perennial herb native to Eurasia and grows up to 1.5m tall. The leaves are simple, lobed and spined, and it produces bright yellow flowerheads 3-5cm wide. The young leaves when only a few inches long can be eaten raw, older leaves can be boiled before eating.

Field sow thistle (Sonchus arvensis)

2469 species of dandelions, native to Eurasia and North America. In general, the leaves are 50–250 mm long or longer, simple, lobed, and form a basal rosette above the central taproot. The flower heads are yellow to orange coloured, and are open in the daytime, but closed at night and during bad weather. The heads are borne singly on a hollow stem that is usually leafless and rises 10–100 mm or more above the leaves. Stems and leaves exude a white, milky latex when broken. The flower heads are 20–50 mm diameter and consist entirely of ray florets. The flower heads mature into spherical seed heads sometimes called blowballs or clocks.

Dandelion, *T. officinalis.* Perennial herb native to Eurasia but found in most temperate regions. It grows from taproots and produces several hollow, leafless flower stems that are typically 5–40 centimetres tall. The leaves are 5-45 cm long and 1-10cm wide and narrow toward the petiole. The leaves are obovate and can be varying degrees of lobed and/or toothed. Each stem has only one flower head, bright yellow and full of strap shaped florets. Leaves, flowers, and roots are all edible.

Dandelion (Taraxacum officinalis)

1 species only in the genus. See below:

Colt's foot, *T. farfara.* Perennial herb native to Eurasia grows up to 30cm tall. The leaves of coltsfoot, which appear only after the flowers have set seed, wither and die in the early summer. The leaves are heart-shaped, with toothed edges and a waxy feel. The flowers superficially resemble dandelions, but the stems have scale leaves. The flower heads are of yellow florets with an outer row of bracts. It is commonly used as a herbal treatment for coughs, but makes a tasty tea too.

Coltsfoot (Tussilago farfara)

CHAPTER 12 – *ROSACEAE*

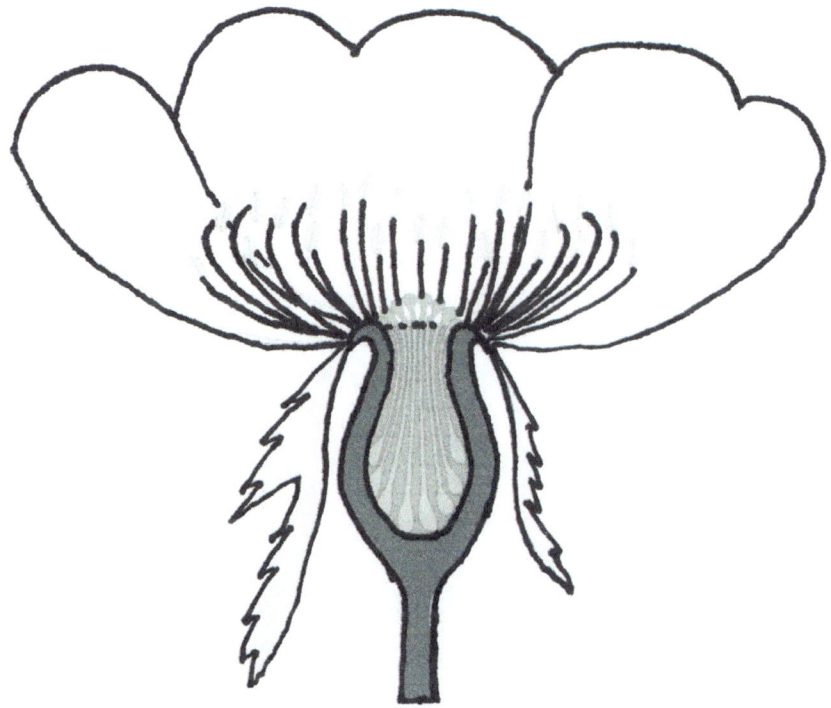

Dog rose (Rosa canina) line drawing

The Rosaceae family, also known as the rose family, is botanically speaking medium sized, but for foragers it's a large one. The family includes herbs, shrubs, climbers, and trees, mostly deciduous but some are evergreen. The herbs are mostly perennials, but a few annuals also exist. The leaf edges are most often serrate, and pared stipules are usually present. The flowers are generally described as "showy". They are radially symmetrical, and almost always hermaphroditic. Rosaceae generally have five sepals, five petals, and many spirally arranged stamens. Solitary flowers are rare. Many rose family plants have edible fruits, but the seeds often contain amygdalin which can convert to cyanide during digestion if the seed is damaged. Rosaceae are found all over the world except Antarctica. The family includes around 5000 species in 108 genera.

CRATAEGUS

224 species of hawthorn, native to temperate regions of Europe, Asia, north America and north Africa. They are shrubs and small trees growing up to 15m tall with small pome fruit and usually thorny branches.

Hawthorn, *C. monogyna.* Shrub or small tree up to 10m tall, and native to Europe, northwest Africa, and west Asia. It has dull brown barks, and the vertical fissures have an orange-ish tinge. The young stems have vicious thorns up to 15mm long. The leaves are up to 40mm long and deeply lobed, dark green on the surface, paler underneath. The inflorescences appear in late spring in corymbs of up to 25 flowers. The flowers are white or pink, multiple stamens, and have a distinctly sweet smell. The pome fruit, also known as a Haw, are small, oval, and dark red, and grow up to 10mm long. The haws always contain a single seed. The young leaves are edible raw and cooked, and the ripe haw flesh is edible.

Common hawthorn (Crataegus monogyna)

13 species of perennial herbaceous plants native to Europe and north America. They tend to grow up to 2m tall with large creamy-white to pink coloured inflorescences of small five petalled flowers.

Meadowsweet, *F. ulmaria.* Perennial herb growing up to 2m tall and native to Europe and west Asia. The stems are tall and upright, furrowed, and sometimes with a hint of pink or purple. The leaves are divided/pinnate with a few pairs of large, serrated leaflets and smaller pairs in between. The terminal leaflets are up to 8cm long and three to five lobed. The leaves are dark green on top and white with soft hairs underneath. The inflorescences are creamy-white and clustered close together. The flowers are small and many with 5 sepals, 5 petals, and up to 20 stamens. The flowers and the crushed foliage have a strong smell which reminds me of TCP antiseptic. The young leaves are used to flavour desserts like custard or cream and can also be used to flavour wine and beer. I find the flowers' flavour too strong for using in food.

Meadowsweet (Fillipendula ulmaria)

24 species of strawberries native to all temperate regions of the world. They are perennial herbaceous plants and tend to grow flat or in clumps, not erect.

Wild strawberry, *F. vesca.* Perennial herb growing up to 15cm tall and native to most of the northern hemisphere. The light green leaves grow from the root on long petioles in a seemingly random spreading pattern. The leaves are trifoliate with toothed edges. The hairy flower stalk is up to 15cm tall which lifts it clear of the foliage where it shows up to 10 white, hairy flowers with 5 sepals and 5 petals. The fruit are small, bright red strawberries. Whilst small, the fruit are delicious, and in my opinion much tastier than their bigger, commercial relatives.

Wild strawberry (Fragaria vesca)

56 species of avens native to most of the northern hemisphere and south America. Closely related to *Fragaria* and *Potentilla*, they have a basal rosette of leaves and produce flowers on thin stalks.

Herb bennet, *G. urbanum.* Also known as wood avens, and clove-root, this perennial herb can grow up to 60cm tall and prefers shady places and is native to the temperate regions of Eurasia and north America. The basal leaves can vary in shape depending on their location and conditions, are pinnate with two to three pairs of leaflets up to 10mm long and one large terminal leaflet. The leaves on the stem are trifoliate, having three leaflets. The bright yellow flowers are 1-2cm in diameter with 5 petals and 5 sepals. The seed head, whilst tiny in comparison, has the same hooked appearance as burdock (*Arctium lappa*). The leaves have a slightly spicy flavour raw, but the roots (as the name suggests) smell and taste of cloves. In cooking they impart the clove flavour and dissolve away.

Wood Avens (Geum urbanum) flower and seed head

36 species of apple trees or shrubs native to the temperate zone of the northern hemisphere. Typically, up to 12m tall with a dense crown. Leaves are up to 10cm long, alternate, simple and with a serrated edge. The flowers grow in corymbs and have 5 petals. The fruit is a globe shaped pome up to 8cm in diameter.

Crab apple, *M. sylvestris.* Deciduous small to medium sized tree can grow up to 14m and live up to 100 years and is native to Europe and western Asia. The trunk diameter is usually up to 45cm, and the bark is light brown and flaky. The leaves are round to oval and sometimes hairy on the underside. The flowers appear in May, usually before hawthorn (*Crataegus monogyna*) and have white, sometimes pale pink petals. The pomes are up to 3cm in diameter. They ripen and fall from the tree in Autumn. Cultivated apples have a definite and visible core, whereas crab apples do not. The fruit can be infused into spirits and fermented as wine. One of my favourite uses is to slice them very thin and dehydrate them into crispy treats.

Crab apple (Malus sylvestris)

64 species of silverweeds, native to the northern hemisphere. They can be annual, biennial, or perennial herbs.

Silverweed, *A. anserina.* Also known as *Potentilla anserina*, this perennial herb likes to grow on riverbanks, meadows and road sides, and is native to the temperate northern hemisphere. It is a low-growing plant with leaves up to 20cm long and evenly pinnate. The leaflets are up to 5cm long with saw-toothed edges. They are green on top and pale grey underneath covered with silky white hairs. They produce single bright yellow flowers on 15cm long stems. They have 5, rarely 7 petals. The roots are eaten as carbohydrate rich, potato-like vegetables.

Silverweed (Potentilla anserina)

340 species of stone fruit producing trees and shrubs, native to north and south America, Eurasia and parts of Africa. They can be deciduous or evergreen. The leaves are simple or alternate. The flowers are white, pink or red and have 5 petals and 5 sepals.

Wild cherry, *P. avium*. Deciduous tree growing up to 18m and native to Europe, west Asia and north Africa. It can have a trunk of up to 1.5m diameter with dark purplish-brown bark with prominent horizontal lenticels. The leaves are alternative, simple, oval up to 14cm long and a serrated edge. Usually matt green on top with fine soft hair underneath. The have a green or reddish petiole with small red glands. The flowers appear in early spring at the same time as the leaves. They are in corymbs with each flower up to 3.5cm wide, five white petals, and yellow stamens. The fruit is a bright red drupe up to 2cm in diameter. The fruit is slightly astringent to bitter raw, but cooks really nicely and infuses into spirits well.

Wild cherry (Prunus avium)

Cherry plum, *P. cerasifera*. Deciduous tree or large shrub growing up to 12m tall, native to southeast Europe and west Asia, and naturalised in the UK. It has hairless, ovate leaves up to 7cm long. It is one of the first trees to flower in spring, often before the leaves are out. The flowers are white or pale pink and up to 2cm across with 5 petals and multiple stamens. The fruit is a drupe up to 3cm in diameter ripening through yellow to red from July to September. It is considered ornamental by many, but it is very fleshy and tastes amazing. One of my favourites.

Cherry plum (Prunus cerasifera)

Blackthorn, *P. spinosa.* Deciduous shrub or small tree growing up to 5m tall and native to Europe, west Asia, and parts of northwest Africa. It has blackish bark and dense, viciously spiky branches. The leaves are oval and up to 4.5cm long with a serrated edge. The flowers are about 1.5cm in diameter with five white petals, and it flowers before the leaves come out in early spring. The fruit is a small black drupe up to 15mm diameter with a purple-blue waxy bloom. They have a large stone and thin, greenish flesh which is very astringent and sour, almost inedible raw. However, cooked or infused in booze they taste amazing and are well worth the effort.

Blackthorn (Prunus spinosa)

267 species of woody, perennial roses native to Asia, Europe, north America, and northwest Africa. They have alternate leaves up to 15cm long, pinnate and 5 to 9 leaflets. The leaflets usually have a serrated edge and prickles on the underside. Most flowers have 5 petals, each petal divided into 2 lobes, and they have 5 sepals. The fruit are red rosehips. The stems have sharp growths called prickles.

Dog-rose, *R. canina*. Deciduous climbing shrub up to 5m tall, although can grow higher climbing up trees and structures, native to Europe, northwest Africa, and western Asia. It is covered in hooked prickles to help it to climb. The leaves are pinnate with 5-7 leaflets with serrate edges. Dog roses flower around June and July with white and pink flowers up to 6cm in diameter with 5 petals and 5 sepals. The red rosehips are up to 2cm long oval fruit. The fruit have thin flesh surrounding multiple hairs and seeds. The hairs are irritant and must not be eaten. The flesh is like a citrus fruit the like of which you've never tasted. It makes a lovely syrup and alcohol infusion.

Dog rose (Rosa canina)

1475 species of brambles shrubs with near cosmopolitan distribution. They typically have woody stems with prickles. They grow long shoots which set roots where they touch the ground. The fruits are aggregate fruits formed from multiple drupelets.

Bramble, *R. fruticosus.* Perennial, woody shrub growing along the ground with stems many metres long and native to most of the world. The stems flower and bear fruits in their second year. In the first year the stem produces large palmately compound leaves with 5 or 7 leaflets. In the second year the stems produce smaller leaves with 3 or 5 leaflets. The flowers appear in late spring on short racemes, each flower about 3cm in diameter with 5 white or pale pink petals. The fruit are black aggregate fruit which everyone knows as a wild treat straight from the bush. Also, young unopened leaf buds taste great, and young shoots can be pickled or candied.

Bramble (Rubus fruticosus)

104 species of whitebeam, rowan, mountain-ash, or service tree shrubs and trees native to the northern hemisphere.

Rowan, *S. aucuparia.* Small deciduous trees up to 12m tall native to northern Europe and mountainous regions of southern Europe and southwest Asia. The tree has a slim trunk and smooth greyish bark. Horizontal lenticels are elongated and orange-brown colour. The leaves are alternate and pinnate up to 20cm long with 4-9 pairs of leaflets with a terminal leaflet. The leaflets are up to 6cm long with serrated edges. The upper side is dark green, and the underside is silvery green. The flowers appear from May to June in large yellow-white corymbs of around 250 flowers. The flowers are up to 10mm and have 5 oval, yellow-white petals with up to 25 stamens, and 5 yellow-green sepals. The flowers and buds have a distinctive marzipan smell. The berries are round, orange pomes up to 10mm in diameter. The fruit are bitter and sour and make an essential part of hedgerow jelly. The flower buds can be picked and coated in chocolate.

Rowan (Sorbus aucuparius)

CHAPTER 13 - *APIACEAE*

Wild carrot (Daucus carota) line drawing

The *Apiaceae* family has 447 genera of mostly aromatic herbs and some shrubs. For foragers and commercial reasons, it is a very important family as it contains many edible species; However, it also contains a few deadly species too. For that reason, I usually advise novice foragers to stay away until they are a bit mor experienced. I'm assuming that as you're reading a book about botany, that you're not a novice so I won't waste our time with repeated warnings.

Common features include that the leaves are always alternate, and mostly emit an aromatic or fetid smell. The main common feature is the inflorescence with the flowers almost always appearing in terminal umbels, hence the older name of the family *Umbelliferae*; They are still sometimes known as the Umbellifers. The flowers have 5 petals and 5 stamens.

The family is also known as the celery, carrot, or parsley family.

CONIUM

6 species of poisonous biennial herbs native to Europe, north Africa, and western Asia. They form basal rosettes in the first year and a hollow flower stalk in the second. They grow best in wet, poorly drained soil rich in nitrogen. They have alternate leaves, pinnately compound and finely divided. The flowers are grouped together in compound terminal umbels.

Hemlock, *C. maculatum.* Biennial herbaceous plant growing up to 2.4m tall and native to Europe and north Africa. It has a hairless, green, hollow stem usually with spots or streaks of purple. The leaves are 2-4 pinnate, finely divided and lacy, overall triangular shaped, up to 50 centimetres long. The flowers are small with 5 white petals, and loosely clustered in umbels. Every part of the plant is toxic and the stems can stay deadly up to 3 years after the plant has died. One of the most dangerous aspects of the plant is how similar it looks to edible species such as cow parsley (*Anthriscus sylvestris*). Hemlock has a distinctive pungent and fetid smell which alone should be enough to prevent ingestion.

Hemlock (Conium maculatum)

34 species of water dropworts, native to much of Eurasia, and parts of Africa and the west coast of north America. Several of the species are very toxic. The small white flowers are arranged in umbellifers.

Hemlock water-dropwort, *O. crocata*. Perennial herbaceous plant growing up to 150cm tall and native to Europe, Asia, and north Africa. It prefers growing in wet or marshy conditions. It has branched, hollow, grooved stems. The upper section of the roots includes 5 or more yellow fleshed obovoid fleshy tubers; Sometimes when it grows on riverbanks and the soil is washed away it exposes the dirty roots from which it gets the name "dead man's fingers". Lower leaves are 3–4 times pinnate, triangular, with oval toothed leaflets 10–20 mm long. The upper leaves are 1–2 pinnate, with narrower lobes and a shorter petiole (leaf stalk). The leaves have a characteristically deceptive smell of parsley or celery. It has compound flower umbels with 12-40 rays which have 5 bracts beneath each. Another deadly poisonous plant with recorded incidents of death caused by eating foliage (similar looking and smelling to celery) and the roots (confused with wild parsnips).

Hemlock Water Dropwort (Oenanthe crocata)

12 species of herbs with compound umbel inflorescences, native to temperate Eurasia.

Ground elder, *A. podograria*. Perennial herb growing up to 100cm native to Eurasia. It has upright, hollow, and grooved stems. The upper leaves are trifoliate, broad, and grooved and the plant has a spreading habit covering wide patches of ground. It flowers in spring and early summer in white petalled compound umbels with 15-20 rays. The young leaves are used as a spring vegetable and potherb, I like to mix it in with pasta, mashed potatoes, and risotto.

Ground elder (Aegopodium podograria)

104 species of tall growing herbs native to temperate and sub-artic regions of the northern hemisphere. They grow to 1–3m tall, with large bipinnate leaves and large compound umbels of white or greenish-white flowers.

Wild angelica, *A. sylvestris*. Annual or short-live perennial growing up to 2.5m tall and native to Europe and central Asia. It has a hollow, hairy stem with purplish furrows. Leaf and flower stems grow from papery sheaths on the main stem. The lower leaves are 3 pinnate and the upper leaves usually 2 pinnate. The flowers are small and white, pink, or purple, borne on umbels in late summer. Traditionally the stems were candied and used to decorate cakes. The young leaves and stems can be boiled and eaten as vegetables.

Wild angelica (Angelica sylvestris)

14 species of chervils native to Europe and temperate Asia. They grow in meadows and verges with wet, porous soil. The hollow stem is erect and branched, ending in compound umbels of small white or greenish flowers. The leaves are bipinnate or tripinnate.

Cow parsley, *A. sylvestris.* Perennial herb, growing up to 170cm tall and native to Europe and Asia. The stems are hollow, furrowed, green sometimes with flushes of purple, and with tiny hairs. The leaf stems have a deep grooved channel with a V-shaped cross section. The leaves are triangular, 2–3 pinnate, up to 45cm long, green, and fern-like or feathery looking. Each flower has 5 white petals, 2 stamens and is arranged in uncrowded compound umbels. The umblets have oval bracteoles with red pointy tips. It is exceptionally similar looking to hemlock (*Conium maculatum*) but smells of parsley, unlike the pungent, musty smell of hemlock.

Cow parsley (Anthriscus sylvestris)

8 species of pignuts native to northwest Europe and the Mediterranean.

Pignut, *C. majus.* Perennial herb growing up to 40cm high and is native to western Europe, the British Isles and Norway. It has a smooth stem, pinnate leaves, and small white flowers in terminal compound umbels. Most well known for its small brown tuber up to 25mm in diameter, which is sweet and aromatic. They really do taste good as a snack, but unless you have a field full of them and plenty of time spare, they're never much more than a snack; Worth the effort though.

Pignut (Conopodium majus) flowers and cleaned tuber

45 species of mostly biennial herbs with some perennial and annuals, found on every continent except antarctica.

Wild carrot, *D. carota.* Variably biennial herb that grows up to 120cm tall. It has rough hairs and a stiff stem. The leaves are bristly, alternate, tripinnate, finely divided and lacy, overall triangular, and up to 15cm long. The small white flowers are clustered in terminal, flat, dense umbels. It is common for the central flower of the umbel to be red. It has 3-forked, or pinnate bracts. As the seeds develop, the umbel curls up and forms a distinctive bird's nest type of structure. The root is edible but very small when young and becomes woody and inedible with age. The seeds can be used as carrot flavoured seasoning.

Wild carrot (Daucus carota) seed head

5 species of fennels native to southern Eurasia and northern Africa.

Wild fennel, *F. vulgare.* Perennial herb growing up to 2.5m tall and native to the mediterranean region. It has a hollow, erect, and green stem. The leaves grow up to 40cm long, finely dissected, with the ultimate segments threadlike. The tiny yellow flowers are borne on compound umbels. The whole plant smells strongly of aniseed. The bulb, leaves and fruit/seeds are used in cooking, although the bulbs of wild fennel aren't as big as cultivated ones. My favourite use is to infuse in vodka.

Wild fennel (Foeniculum valgare)

90 species of hogweeds or cow parsnips native throughout the temperate northern hemisphere.

Hogweed, *H. sphodylium.* Perennial herb growing up to 2m tall and native to most of Europe, western Asia, and northern Africa. It has hollow, ridged stem with bristly hairs coming from a large tap root. The leaves can reach 50cm long. They are once or twice pinnate, hairy and serrated, divided into 3–5 lobed segments. The leaves appear from papery sheaths considered to be a delicacy. Each flower has five white or pinkish-white to purplish petals and is arranged in large umbels. The seeds, both green and young, and brown and older are also edible and commonly used as a spice; It reminds me of cardamom. When the plant is young it can look very similar to giant hogweed (*Heracleum mantegazzianum*) which contains phototoxic compounds.

Common hogweed (Heracleum sphondylium)

5 species of alexanders native to the mediterranean and naturalised in Britain.

Alexanders, *S. olusatrum.* Biennial herb growing up to 180cm tall, native to the mediterranean and naturalised in Britain. It is a hairless plant with a solid stem up to 22 mm in diameter, which becomes hollow and grooved with age. It has a tuberous taproot which can be 60 cm long. The stem leaves are arranged in a spiral with an inflated, purple-striped, fleshy petiole that has papery margins towards the base. The compound leaves are diamond-shaped, 2- or 3-times ternately (sometimes pinnately) divided. The individual leaflets are dark green above, pale green below, flat, lobed, and serrated with obtuse teeth that have a tiny white tip. Inflorescences are terminal or in the leaf axils, arranged in compound umbels up to 10cm in diameter. The flowers are small, with 5 yellowish petals and 5 tiny, green sepals, and 5 stamens. The fruit are a black hard nut-like schizocarp with strong smelling oils. The young plants have been eaten as herbs and vegetables, but I find the flavour a little too pungent. However, the mature fruit can be ground and used as a spice quite reminiscent of Szechuan pepper.

Alexanders (Smyrnium olusatrum)

AFTERWORD

I just wanted to close this book by saying that foraging, fieldcraft, herbalism and botany are all passions of mine, and I will gladly talk at people about them for hours on end; However, I genuinely believe that a little bit of botany will benefit the developing forager greatly.

It may be that upon reading this book, you think I've gone too deep, or maybe you think I haven't gone deep enough. Honestly, it was a challenge to not just write everything I've learned about botany over the years, but I think I got the balance about right. There may well be new version in the coming years so if you think something is missing, or some content is of no use, please email me and let me know and I'll consider your suggestions for the next edition.

Finally, I'd just like to say a big thank you to the foragers whom I've spoken to, taught, and learned from over the years that I've been writing this book. You may not have directly contributed to the content, but you've inspired me and kept me going all this time.

Glossary

I've tried to explain botanical terminology as I go with this book, that is kind of the point after all, but just in case you don't want to go through the whole book to look something up, I've also tried to include them all here too.

Please note that this is not an exhaustive list. There are many more botanical terms, but I've tried to keep it to the ones that a forager or wildcrafter might come across/care about.

A

- Aerial: The parts of a plant growing above the ground.
- Aggregate fruit: Clusters of fruits from multiple carpels of a single flower.
- Alternate: Arrangement where leaves grow offset from each other on the stem.
- Annual: Plants that complete their lifecycle in one year, then die.
- Anther: The pollen producing part of a flower.
- Apex: The tip of a leaf.
- Aril: Fleshy, berry-like fruit which doesn't entirely cover the seed.
- Aromatic: Producing volatile oils detectable by smell.
- Axil: The upper angle of the node where the petiole joins the stem.

B

- Basal rosette: Leaf arrangement where leaves spread out from a central point along the ground.
- Berry: Fruit in which the seeds are contained in the fleshy pulp.
- Biennial: Plants that complete their lifecycle in two years, then die.
- Bifoliate: Two leaflets growing together.
- Binomial: The scientific naming convention of all life on Earth. Literally means two names.
- Bipinnate: A leaf which divides into two or more stems, each containing a pinnate arrangement of leaflets.
- Blade: The main part of the leaf a.k.a. Lamina.
- Bract: A modified type of leaf, usually associated with flowers.

- Bulb: A thickened underground energy storage part consisting of stem and leaf bases.
- Burr: A prickly fruit with hooks to attach to fur/clothes.

C

- Calyx: The collective part of a flower which includes all sepals.
- Capitulum: Dense cluster of florets, such as in daisy family.
- Carpel: The female part of a flower including ovary, style and stigma.
- Cordate: Heart shaped leaf.
- Corm: Underground, fleshy, swollen base with buds or covered by thin scales.
- Corolla: The collective part of a flower which includes all petals.
- Crenate: Rounded teeth leaf edge.
- Cyme/cymose: Arrangement of flowers where all flower stems and branches end in a flower.

D

- Deciduous: Drops leaves seasonally.
- Dehiscent: Dry fruit which break open at maturity to release their seeds.
- Dentate: Symmetrical teeth shaped leaf edge.
- Digitate: Arrangement of leaflets spreading out from one point at the tip of a stem a.k.a. Palmate.
- Dioecious: Plant which has separate male and female flowering plants.
- Drupe: Type of fleshy fruit that has a single seed inside a hard wall.
- Drupelet: Small drupes which cluster together forming a compound fruit.

E

- Entire: Even and smooth leaf edge.
- Erect: Grows upwards and is self-supporting.
- Even pinnate: Two or more pairs of opposite arranged leaflets forming a leaf.
- Evergreen: Keeps its leaves all year round.

F

- Family: The highest level of classification that we use.

- Filament: The part of a flower which supports the anther.
- Floret: Small flower found in large clusters.
- Flower: The sexual reproductive part of a flowering plant.

G

- Gall: Abnormal growth caused by viruses, fungi, bacteria, insects, and mites.
- Genus (pl. genera): The second lowest level of classification that we use. When combined with species, usually enough to identify a specific plant. The first part of binomial scientific names (Genus species).
- Glabrous: Smooth, lacking hairs.

H

- Herbaceous: A plant which does not develop a woody stem.
- Hip: The fruit of a rose plant.
- Hirsute: Hairy.

I

- Immaculate: Not spotted.
- Incomplete flower: A flower which grows without one or more of the usual parts, such as sepals, petals, stamens etc.
- Indehiscent: Dry fruit which remain closed at maturity.
- Inflorescence: Arrangement of flowers.
- Internode: The space between nodes.

K

- Key (1): A plant reference which helps you to narrow down the identification of a plant.
- Key (2): A type of fruit/seed produced for plant reproduction (e.g. Ash keys).

L

- Lamina: Main part of leaf a.k.a. Blade.
- Lanceolate: Long, point-ended leaf.
- Latex: A milky sap which exudes from some plants and fungi.
- Leaf: Can be made up of one or more leaflets.
- Leaflet: Arrangement of leaf-like structures that together form a leaf.

- Lenticel: Typically lens-shaped and horizontal porous tissue in bark which allows exchange of gases with the atmosphere.
- Linear: Long with parallel sides shaped leaf.
- Lobed: Deeply indented leaf edges, usually not-quite as deep as the mid-rib.

M

- Maculate: Spotted.
- Margin: The edge of a leaf.
- Midrib: The central vein of a leaf, usually running from base to tip.
- Monoecious: Plant which has both male and female flowers on one plant.

N

- Native: Historically occurring in a specific area.
- Naturalised: Not originally from the area but introduced a long time ago.
- Node: Where petiole's attach to the plant stem.
- Nomenclature: Literally means name.
- Nut: A hard, dry fruit containing one seed.

O

- Ob-: When put before a leaf shape, means the other way around. E.g. Obovate, meaning egg-shaped with the narrower end near the stem.
- Odd pinnate: Two or more pairs of alternately arranged leaflets, usually with a terminal leaflet at the end.
- Opposite: Arrangement where leaves grow opposite each other on the stem.
- Ovary: The female part of the flower which produces ovules.
- Ovate: Egg shapcd leaf.

P

- Palmate: Arrangement of leaflets spreading out from one point at the tip of a stem a.k.a. Digitate.
- Panicle: A compound raceme flower arrangement.
- Peduncle: The stem of an inflorescence.
- Pendulous: Having a drooping growth habit.
- Perennial: Plants that live for more than two years, usually completing a cycle ending in seed production every year.

- Petal: Usually colourful, showy parts of the flower which open up to reveal the reproductive parts.
- Petiole: Leaf stem.
- Pollen: The male product of flowers, including the sperm cells for plant sexual reproduction.
- Pome: A fleshy fruit of the *Rosaceae* family with a central core containing multiple seeds and surrounded by a tough membrane.
- Pistil: The parts of a flower responsible for producing an ovule (egg).
- Prostrate: Grows along the ground as oppose to erect.
- Pubescent: Covered in downy hair.

R

- Raceme: An inflorescence where the main stem produces flowers on lateral stalks, starting at the base and the last flowers opening at the tip.
- Radial: With structures growing from a central point and spreading outwards.
- Reniform: Kidney shaped.
- Rhizome: An underground energy storage part, usually growing horizontally and persists from year to year.
- Rugose: Wrinkled.

S

- Samara: A dry fruit with a papery wing, such as on Ash or Hornbeam.
- Seed leaves: The first leaf or leaves to appear on seed germination. Not true leaves and sometime called cotyledons.
- Sepal: One of the outer parts of a flower, usually green, and usually surround the petals.
- Serrate: Backward pointing teeth on leaf edge.
- Sessile: Attached without a stalk.
- Shrub: Woody perennial plant without a single main stalk, usually smaller than a tree.
- Species: The lowest level of classification that we use. Usually enough to identify a specific plant. The second part of binomial scientific names.
- sp: Used to refer to a plant when the Genus is known but the species is not.
- spp: Used to refer to every species within a genus. E.g. *Taraxacum* spp.

- Spadix: A spike-like inflorescence with densely crowded flowers.
- Spathe: A large bract which en-sheathes a spadix.
- Stamen: The male parts of a flower including filament and anther.
- Stigma: The female part of a flower which receives the pollen.
- Stipule: A leaf-like structure growing from the leaf axil. Usually much smaller than the leaf.
- Style: The stem of the female part of the flower which delivers the pollen to the ovary.
- Succulent: Juicy or fleshy.

T

- Taproot: A single descending root of a plant.
- Taxonomy: The study of biological classification.
- Terminal: At the end of the stem.
- Trifoliate: Three leaflets growing together.
- Tripinnate: A leaf which divides into two or more stems, each dividing into two or more stems again, each containing a pinnate arrangement of leaflets.
- Tuber: General name for underground energy storage parts that have no other function. Does not include taproots, or corms.

U

- Umbel: A raceme inflorescence type in which all the flower stalks cluster together at the top of the stem with varying stem lengths to produce a flattened inflorescence, or similar lengths to produce a globular inflorescence. The name has the same origin as umbrella, and the collection of stems is similar to the structure of an umbrella.
- Undulate: Wavy.

V

- Variegated: Having irregular patches of another colour.
- Vein: Used to transport nutrients around leaves.

W

- Whorled: Leaf arrangement where 3 or more leaves grow around the stem at the same point.

RECOMMENDED RESOURCES

Obviously, this book is perfect in every way and includes every detail you could ever need, but just in case you're tempted to look elsewhere, here are some of the resources I recommend.

WEBSITES:

- www.foundfood.com
- www.foragerhelper.foundfood.com
- www.botanydepot.com
- www.coursera.org/learn/plantknows
- www.foragers-association.org

Please note that we cannot guarantee that external websites will still be available when you read this book, but they were at the time of writing.

BOOKS:

- Plant Names Simplified – Adrian Stockdale (5M publishing 2019)
- Botany in a Day – Thomas Elpel (HOPS press 2013)
- Identifying and Harvesting Edible and Medicinal Plants – Steve "Wildman" Brill (William Morrow 1994)
- The Botany Colouring Book – Paul Young (Harper 1982)
- What a Plant Knows – Daniel Chamovitz (Farrar, Strauss, and Giroux 2012)

INDEX

ABOUT THE AUTHOR

Gavin is passionate about understanding how plants and fungi live and work and how we can use them to our benefit without causing harm to the environment. Having had a very 'outdoors' childhood in the 1970s, Gavin spent ten years serving in the British Army where he learned about emergency food and first aid (including the beginnings of foraging and herbalism). After a second (third?) career in IT, Gavin reignited his passion for the natural world and started studying foraging in earnest. One thing led to another and before you know it he's studying herbalism and botany too, and running regular walks in and around London to introduce others to the fascinating world all around us.

Gavin formed FoundFood.com in 2012 as a personal database of his learnings, so he would always be able to look up where he had learned something from. In 2017 FoundFood.com became a blog to share some foraging related musings and experiences, them in 2023 Gavin started making his database of foraging information available to subscribers, and it has continued to grow ever since.

Now, in 2024, after many requests, Gavin has begun to release online courses and books available from www.foundfood.com

WS - #0218 - 260825 - C132 - 230/190/7 - PB - 9781836020691 - Matt Lamination